James Parish Stelle

The American Watchmaker and Jeweler

James Parish Stelle

The American Watchmaker and Jeweler

ISBN/EAN: 9783337021399

Printed in Europe, USA, Canada, Australia, Japan

Cover: Foto ©Andreas Hilbeck / pixelio.de

More available books at **www.hansebooks.com**

HANEY'S TRADE MANUALS.

AMERICAN

WATCHMAKER

AND

JEWELER;

A CLEAR AND COMPLETE EXPOSITION

OF ALL

THE LATEST AND MOST APPROVED

Secrets of the Trade,

WITH A SERIES OF

PLAIN INSTRUCTIONS FOR BEGINNERS,

ETC., ETC.

BY J. PARISH STELLE,

A PRACTICAL WATCHMAKER.

New York:

JESSE HANEY & CO., PUBLISHERS,

No. 119 NASSAU STREET.

THE

AMERICAN WATCHMAKER

AND

JEWELER,

A FULL AND COMPREHENSIVE EXPOSITION

OF ALL THE

Latest and most Approved Secrets of the Trade

EMBRACING

WATCH AND CLOCK CLEANING AND REPAIRING, TEMPERING
IN ALL ITS GRADES, MAKING TOOLS, COMPOUND-
ING METALS, SOLDERING, PLATING, ETC.,

WITH A SERIES OF PLAIN INSTRUCTIONS FOR BEGINNERS.

ALSO,

DIRECTIONS BY WHICH THOSE NOT FINDING IT CONVE-
NIENT TO PATRONIZE A HOROLOGIST MAY KEEP
THEIR CLOCKS IN ORDER.

BY J. PARISH STELLE,

A PRACTICAL WATCHMAKER.

New York:

JESSE HANEY & CO., PUBLISHERS,

119 NASSAU STREET.

HANEY'S TRADE MANUALS.

PUBLISHERS' NOTICE.

THE AMERICAN WATCHMAKER AND JEWELER is the initial issue of a series of "TRADE MANUALS" which we propose publishing as fast as they can be properly prepared. The value of books treating of the processes, manipulations and discoveries of the different trades and professions is recognized by every intelligent man. While no book can pretend to be a substitute for experience and natural ability, in the prosecution of any industry, there is unquestionably much information that can be told in a few moments, which might require years to arrive at by individual experiments. A guide of this kind should embrace the combined results of all discoveries and improvements in the art of which it treats, so as to keep pace with the progress constantly taking place.

There are many good books relating to the different trades, and we consider that a liberal expenditure for such would prove profitable to every artisan. Such books are necessarily higher priced than common works. The main objection to most of them is that they contain much unimportant matter which swells them in both size and price, at the same time that it confuses the reader. We shall attempt to obviate this objection in HANEY'S TRADE MANUALS by giving in concise form all the really valuable information attainable on the subjects treated. Great care will be taken to make them reliable in every respect, and of real assistance to the reader. They will be almost wholly original, written by practical and experienced men. In order that they may have a large and general circulation, and be within reach of every person engaged in the occupation they treat of, HANEY'S TRADE MANUALS will be sold at the lowest prices, considering the original cost of preparation, manufacture of books, and the extent of the demand. Those which, from the limited extent of any trade, necessarily have but a small circulation, must of course be charged somewhat higher than those appertaining to more extensive interests. We shall, however, in every case content ourselves with a moderate and reasonable profit in our investment

PREFACE.

I DO not expect all watchmakers to praise me for having presented this exposition of the " secrets of the trade," no more than the physician who produces a work adapted to domestic practice could reasonably expect a blessing from every member of the medical profession. It is all the same to me. I did not write it for praise ; nor did I write it under the conviction that I was doing anything deserving of blame. I expect to *be* blamed, however, and to have hard things said of me by a few who either feel that they know enough already to make out with, and would rather not have the secrets imparted to others lest they in consequence should come in for a share in the success ; or who are making a good thing of it by selling " The Latest and Most Improved Processes " to the less fortunate, at the moderate price of from three to forty dollars each. I know just about what they will say of me : that I shall leave between them and their own consciences. I know, too, about how they will argue with a view to creating a foundation for blame : of that I may speak a word or two. They will intimate that there are too many " botches" in the business already, and that a work of this character is only calculated to augment their number.

At first thought this thing may appear reasonable enough, but a sober second reflection will convince any reasonable person of its inability to hold good. That there are "botches," and sad ones, following the vocation I must admit ; and what is still worse, many of them are men who have enjoyed excellent opportunities for gaining information. As a general thing, a lack of capacity rather than information has made them "botches," and this very custom of husbanding the " secrets of the trade " from the public is what enables them to curse the communities in which they are located, by holding positions which would otherwise be filled by better men. So long as horological information can only be obtained at a high price, a large number of unqualified persons, who happen to be financially favored, will buy it, and " botches" must be the result, of course. But not so when we place it within the reach of all. Men qualified for the business will then take hold of it, and such as are now imposing upon the people, simply because they happen to possess a few secrets too costly to be generally known, will find themselves under the necessity of falling back. France is said to be blessed with the most skillful watchmakers in the world, and the reason is plain—she is the only nation whose authors have attempted to produce a series of cheap and reliable books on the science of horology.

This is my argument. It was what prompted me to write this book ; and though I might produce other arguments in favor of the move, I think it is enough.

I shall not speak of the character or claims of the AMERICAN WATCHMAKER AND JEWELER, preferring that the book should show for itself. A knowledge of the fact that superior opportunities for acquiring the latest and best information touching this, my favorite subject, have presented themselves and been eagerly embraced by me, both in Europe and America, emboldens me to send forth the work without a single misgiving.

THE AUTHOR.

CONTENTS.

————◆◆◆————

INTRODUCTION.

ON WATCH CLEANING.

ON WATCH REPAIRING.

ON MENDING WATCH TRAINS.

ON TEMPERING.

ON PLATING.

MISCELLANEOUS RECIPES.

THE AMERICAN

WATCHMAKER AND JEWELER.

——∘o⫶ᴑ⫶o∘——

CHAPTER I.

INTRODUCTION.

THE American watchmaker, so called, is not usually a
manufacturer of watches, or even parts of watches, but
simply an artist whose business it is to repair and keep
watches in order. He is generally a man of rare mechanical genius, capable of turning his hand to almost anything,
hence he is not unfrequently, especially in the country, also
a clockmaker—in the same sense—a jeweler, and a repairer
of musical instruments. In short the good watchmaker is
almost invariably, if he is disposed to let himself out, a
Jack-of-all-trades. He must possess a degree of ingenuity
sufficient to qualify him for almost any mechanical perform-
ance without the benefit of a previous apprenticeship, or he
cannot be a successful watchmaker, for it is a business in
which there is no regular routine, as in other trades. Any
industrious person, though endowed with nothing above an
ordinary capacity, may, in obedience to a long series of instructions combined with practice, make a master carpenter,
blacksmith or wheelwright of himself, but not a watchmaker.
The watchmaker whose skill is to render him deserving of
the application, must be blessed with a natural gift above
the generality. Like the painter, the sculptor or the poet,
he must be born to the calling. Not only must he be what

is termed a natural mechanic, but a philosopher as well, possessed of a good reasoning power of his own; for instances are sure to occur, and often, in which he will be called upon to ferret out causes and effects never met with or thought of by his instructions.

I throw in these hints, not with a view to the discouragement of any, but in the hope that they may be of benefit to some who are thinking of becoming watchmakers. If the true element is in them it has given evidence of the fact, and they may go ahead with confidence of success; if not, they had better abandon the idea at once and turn attention to something else; bearing in mind that all were not made for the same vocation, and that he who would not make a useful watchmaker, might more than succeed at some other calling. True, a person might get along at the business without these extra qualifications named, but there would be no chances for him to excel, and unless one could be an excellent watchmaker he had far better be no watchmaker at all. Unfortunately for us, and for them, there are already too many second and third class workmen of the kind in America.

Parents who contemplate putting their children to trades should bear in mind the important truths on which I have just been treating. The best years of a boy's life may be literally wasted in the acquirement of a vocation for which he has no natural qualifications.

To within a few years back horology was at a low ebb in the United States. It is beginning to look up now, however, with excellent prospects for a glorious future. I am of the opinion that the day is not far distant when she will make not only all her own time-pieces, but will furnish a very large proportion of those used in other parts of the world. This conclusion I base upon what she has done and is doing already. It is truly astonishing when we take into consideration the fact that the business was a stranger to her shores up to the beginning of the nineteenth century.

The first attempt at producing machines on American soil for the measurement of time was made by Eli Terry of Plymouth Hollow, Conn., A. D. 1800, in the manufacture of the old fashioned wooden clocks. He went into the business on an exceedingly small scale at first, doing, I think, all the work himself, and acting as his own salesman and traveling agent. He would finish two or three clocks, it is said, and swinging them upon the back of a horse, would strike out into the country and peddle till the last one was sold; then, but not till then, he would return to his home and engage in the manufacture of a new cargo.

The excellence of Mr. Terry's clocks, and their cheapness when compared to that of the imported article, soon caused his business to grow until the erection of a large establishment became necessary. This continued in successful operation until Mr. Terry's death a few years ago.

When it became known that the Plymouth Hollow clock factory was a paying institution, other establishments sprung up to rival it. Great improvements were made both in the materials worked and the manner of working them. Indeed, so rapid was the progress made that only a few brief years passed ere America was famed abroad for producing the best clocks in the world, and large exportations were constantly being made.

An establishment for the manufacture of watches went into operation at Worcester, Mass., in 1812, but soon failed. In 1830, another was started at Hartford, Conn., but after turning out near one thousand watches it too went down, and the hope of competing successfully with English work seemed to die out for the present.

In 1850, Mr. A. L. Dennison of Maine suggested the idea of manufacturing a watch entire in one establishment, by properly constructed machinery—a thing not yet thought of in Europe. Others took with the idea and soon joined him in the erection of a manufactory at Roxbury, Mass.

The plan worked to the satisfaction of all concerned, but the site was found to be unsuitable on account of the dust; consequently, in 1854, the concern was removed to Waltham, in the same state, where it is still (1868) in successful operation, turning out the celebrated " American Watches " in large numbers. It is known as " The American Watch Company of Waltham, Mass.," and its watches have acquired a good reputation.

A second watch manufactory on Mr. Dennison's plan, was established at Nashua, New Hampshire, but want of capital soon caused it to fail, and the American Watch

Company bought its machinery. A third is now in operation at Elgin, Illinois, near Chicago, under style of "The
National Watch Company." It was established in 1867,
and its productions have a very excellent reputation.

———o○;○;○○———

CHAPTER II.

ON WATCH CLEANING.

IT is hardly necessary to say that great caution must be
observed in taking the watch down—that is, in separating
its parts. If you are new at the business think before you
act, and then act slowly. Take off the hands carefully so
as not to bend the slender pivots upon which they work;
this will be the first step. Second—loosen and lift the
movement from the case. Third—remove the dial and dial
wheels. Fourth—let down the main-spring by placing your
bench key upon the arbor, or "winding post," and turning
as though you were going to wind the watch until the click
rests lightly upon the ratchet; then with your screw-driver
press the point of the click away from the teeth, and ease
down the springs. Fifth—draw the screws (or pins) and
remove the bridges of the train, or the upper plate, as the
case may be. Sixth—take out the balance. Great care
must be observed in this or you will injure the hair-spring.
The stud or little square post into which the hair-spring is
fastened may be removed from the bridge or plate of most
modern watches, without unkeying the spring, by slipping a
thin instrument, as the edge of a knife blade, under the
corner of it and prizing upward. This will save you a considerable amount of trouble, as you will not have the hair
spring to adjust when you reset the balance.

If the watch upon which you propose to work has an
upper plate, as an American or an English lever for instance,
loosen the lever before you have entirely separated the
plates, otherwise it will hang and most likely be broken.

Having the machine now down, brush the dust from its

different parts and subject them to a careful examination with your eye-glass. Assure yourself that the teeth of the wheels and leaves of the pinions are all perfect and smooth; that the pivots are all straight, round and highly polished; that the holes through which they are to work, are not too large, and have not become oval in shape; that every jewel is smooth and perfectly sound; and that none of them are loose in their settings. See, also, that the escapement is not too deep or too shallow; that the lever or cylinder is perfect; that all the wheels have sufficient play to avoid friction, but not enough to derange their coming together properly; that none of them work against the pillar-plate; that the balance turns horizontally and does not rub; that the hair-spring is not bent or wrongly set so that the coils rub on each other, on the plate or on the balance; in short, that everything about the whole movement is just as reason would teach you it should be. If you find it otherwise, proceed to repair in accordance with a carefully weighed judgment, and the processes given in next chapter, after which clean—if not, the watch only needs to be cleaned, and therefore you may go ahead with your work at once.

TO CLEAN.

Many watchmakers wet the pillar plates and bridges with saliva, and then dipping the brush into pulverized chalk or Spanish whiting, rub vigorously until they appear bright. This is not a good plan, as it tends to remove the plating and roughen the parts, and the chalk gets into the holes and damages them, or sticks around the edges of the wheel-beds. The best process is to simply blow your breath upon the plate or bridge to be cleaned, and then to use your brush with a little prepared chalk—(See recipe for preparing it.) The wheels and bridges should be held between the thumb and finger in a piece of soft paper while undergoing the process; otherwise the oil from the skin will prevent their becoming clean. The pinions may be cleaned by sinking them several times into a piece of pith, and the holes by turning a nicely shaped piece of pivot wood into them, first dry and afterwards oiled a very little with watch oil. When the holes pass through jewels you must work gently to avoid breaking them.

The oiling above named is all the watch will need. A great fault with many watchmakers lies in their use of too much oil.

THE " CHEMICAL PROCESS."

Some watchmakers employ what they call the "Chemical Process" to clean and remove discolorations from watch movements. It is as follows :—

Remove the screws and other steel parts ; then dampen with a solution of oxalic acid and water. Let it remain a few moments, after which immerse in a solution made of one-fourth pound cyanuret potassa to one gallon rain water. Let remain about five minutes, and then rinse well with clean water, after which you may dry in sawdust, or with a brush and prepared chalk, as suits your convenience. This gives the work an excellent appearance, but I cannot say that it makes it any better than does the old process.

TO PREPARE CHALK FOR CLEANING.

Pulverize your chalk thoroughly, and then mix it with clear rain water in the proportion of two pounds to the gallon. Stir well and then let stand about two minutes. In this time the gritty matter will have settled to the bottom. Pour the water into another vessel, slowly so as not to stir up the settlings. Let stand until entirely settled, and then pour off as before. The settlings in the second vessel will be your prepared chalk, ready for use as soon as dried.

Spanish whiting treated in the same way makes a very good cleaning or polishing powder. Some operatives add a little jeweler's rouge, and I think it is an improvement ; it gives the powder a nice color at least, and therefore adds to its importance in the eyes of the uninitiated. In cases where a sharper polishing powder is required, it may be prepared in the same way from rotten stone.

PIVOT WOOD.

Watchmakers usually buy this article of watch-material dealers. A small shrub known as Indian arrow-wood, to be met with in the northern and western states, makes an excellent pivot wood. It must be cut when the sap is down, and split into quarters so as to throw the pith outside of the rod.

PITH FOR CLEANING.

The stalk of the common mullen—*verbascum thapsus*—affords the best pith for cleaning pinions that I have ever yet tried. It may be found in old fields and by-places all over the country. Winter, when the stalk is dry, is the time to gather it. Some workmen use cork instead of pith, but it is not so good and far less safe.

CHAPTER III.

ON WATCH REPAIRING.

I SHALL not attempt to describe, and to prescribe for, every species of defect that has been known to occur in a watch, for two reasons : The first is, that it would make a work far too large to come within the scope of my present plans, or to be useful ; and the second, that many of the defects constantly to be met with are of a character so simple, and so plain, that any person with ordinary ingenuity will be able to note them at once and apply the remedy. Such, for instance, as putting in a main-spring, a hair-spring, or a jewel; a mere glance at the machine will be sufficient to satisfy the proposed operative with regard to the steps necessary to be taken, even though he may have never before seen the inside of a watch.

With a view, then, to giving my reader the largest possible amount of useful information for his money, I shall proceed at once to offer such modes employed in watch repairing as he could not easily acquire himself—in short, to present in the briefest possible manner a complete exposition of those processes in use, known as " The Secrets of the Trade." Once they are mastered, he will find it no longer a difficult matter to carry on the watch-repairing business with credit and success ; provided, of course, he possesses a reasonable amount of ingenuity and patience.

TO PIVOT.

When you find a pivot broken, you will hardly be at a loss to understand that the easiest mode of repairing the damage is to drill into the end of the pinion or staff, as the case may be, and having inserted a new pivot, turn it down to the proper proportions. This is by no means a difficult thing when the piece to be drilled is not too hard, or when the temper may be slightly drawn without injury to the other parts of the article. It will be difficult, however, in cases where you find it necessary.

TO DRILL INTO HARDENED STEEL.

For this purpose make your drill oval in form, instead of in the usual shape, and temper as hard as it will bear without crumbling. Roughen the surface of the object into which you desire to drill with a little diluted nitric acid. Start your drill, and to prevent it from becoming heated use spirits of turpentine instead of oil. Some workmen use kerosene with gum camphor dissolved in it instead of turpentine.

When your drill begins to run smooth in consequence of the bottom of the holes becoming burnished, clean out the turpentine or kerosene and roughen again with acid; then proceed as before.

You will find this a somewhat tedious business, but with a little patient application you will finally be able to accomplish your object. It is the only mode for drilling into highly tempered steel that will work with any degree of certainty.

TO TELL WHEN THE LEVER IS OF PROPER LENGTH.

You may readily learn whether or not a lever is of proper length, by measuring from the guard point to the pallet staff, and then comparing with the roller or ruby-pin table; the diameter of the table should always be just half the length measured on the lever. The rule will work both ways, and may be useful in cases when a new ruby-pin table has to be supplied.

TO LENGTHEN LEVERS OF ANCHOR ESCAPEMENTS.

Some do this by drawing the temper of the lever between the pallets and the fork and forging it out to the proper length ; others by soldering a piece the required thickness against the guard point just back of the fork.

There is a new process advertised by dealers in the "Secrets of the Trade"—price three dollars—as " The best and quickest means of bringing the point of the lever close to the roller, without hammering the point, soldering a piece on or stretching the lever." It is as follows :—

Cut across with a screw-head file, just back of the fork, as deeply as you can with safety. The thin point thus left standing to itself you will bend gently forward to the proper position. This is all that will be required. In the event you break the little point in your efforts to bend it—a thing not likely to happen—you can file down level, drill a hole and insert a pin American lever style.

TO CHANGE DEPTH OF LEVER ESCAPEMENT.

If you are operating on a fine watch the best plan is to put a new staff into the lever, cutting its pivots a little to one side—just as far as you desire to change the escapement. Common watches will not, of course, justify so much trouble. The usual process in their case is to knock out the staff, and with a small file cut the hole oblong in a direction opposite to that in which you desire to move your pallets ; then replace the staff, wedge it to the required position, and secure by soft soldering.

In instances where the staff is put in with a screw you will have to proceed differently. Take out the staff, prize the pallets from the lever, file the pin holes to slant in the direction you would move the pallets, without changing their size on the other side of the lever. Connect the pieces as they were before, and with the lever resting on some solid substance you may strike lightly with your hammer until the bending of the pins will allow the pallets to pass into position.

TO TELL WHEN THE LEVER PALLETS ARE OF PROPER SIZE.

The clear space between the pallets should correspond

with the outside measure, on the points, of three teeth of the scape wheel. The usual mode of measuring for new pallets is to set the wheel as close as possible to free itself when in motion. You can arrange it in your depthing tool, after which a measurement between the pivot holes of the two pieces, on the pillar plate, will show you exactly what is required.

TO PUT TEETH INTO WHEELS

Most watchmakers solder or dovetail their teeth in, but there is a new mode which I consider far better, and I know it is easier: Make a hole through the plate of the wheel immediately below the point from which the tooth has been broken. Let its diameter be a little greater than the width of a tooth. Next, with your tooth-saw cut down where the tooth should stand till you come into the hole. You then dress out with a head upon it, a piece of brass wire, till it fits nicely into the cut of the saw, with its head in the hole. With a fine graver you then cut a crease into the wheel-plate above and below, on either side of the newly-fitted wire; after which, with your hammer, you cautiously spread the face of the wire until it fills the creases, and is securely clinched or riveted into the wheel. This makes a strong job, and one that dresses up to look as well as any other.

TO WEAKEN THE HAIR-SPRING.

This is often effected by grinding the spring down. You remove the spring from the collet, and place it upon a piece of pivot wood cut to fit the centre coil. A piece of soft steel wire, flattened so as to pass freely between the coils, and armed with a little pulverized oil-stone and oil, will serve as your grinder, and with it you may soon reduce the strength of the spring. Your operations will, of course, be confined to the centre coil, for no other part of the spring will rest sufficiently against the wood to enable you to grind it, but this will generally suffice. The effect will be more rapid than one would suppose, therefore it will stand you in hand to be careful or you may get the spring too weak before you suspect it.

Another and perhaps later process is as follows : Fit the

collet, without removing the spring, upon a stick of pivot-wood, and having prepared a little diluted nitric acid in a watch-glass, plunge the centre coils into it, keeping the other parts of the spring from contact by holding it in the shape of an inverted hoop skirt, with your tweezers. Expose it a few seconds, governing the time of course by the degree of effect desired, and then rinse off, first with clean water, and afterwards with alcohol. Dry in the sun or with tissue paper.

TO PREVENT A CHAIN FROM RUNNING OFF THE FUSEE.

In the first place you must look after and ascertain the cause of the difficulty. If it results from the chain's being too large, the only remedy is a new chain. If it is not too large, and yet runs off without any apparent cause, change it end for end—that will generally make it go all right. In cases where the channel in the fusee has been damaged, and is rough, you will be under the necessity of dressing it over with a file the proper size and shape. Sometimes you find the chain naturally inclined to work away from the body of the fusee. The best way to remedy a difficulty of this kind is to file off a very little from the outer lower edge of the chain the entire length—this, as you can see, will incline it to work on instead of off. Some workmen, when they have a bad case, and a common watch, change the standing of the fusee so as to cause the winding end of its arbor to incline a little from the barrel. This, of course, cannot do otherwise than make the chain run to its place.

TO PUT WATCHES IN BEAT.

If a cylinder escapement, or a detached lever, put the balance into position, then turn the regulator so that it will point directly to the pivot-hole of the pallet staff if a lever, or of the scape-wheel if a cylinder. Then lift out the balance with its bridge or clock, turn it over and set the ruby pin directly in line with the regulator, or the square cut of the cylinder at right angles with it. Your watch will then be in perfect beat.

In case of an American or an English lever, when the regulator is placed upon the plate, you will have to proceed differently. Fix the balance into its place, cut off the connection of the train, if the mainspring is not entirely down, by slipping a fine broach into one of the wheels, then look between the plates and ascertain how the lever stands. If the end farthest from the balance is equi-distant between the two brass pins it is all right—if not, change the hairspring till it becomes so.

If dealing with a duplex watch, you must see that the roller notch, when the balance is at rest, is exactly between the locking tooth and the line of centre—that is, a line drawn from the centre of the roller to the centre of the scape-wheel. The balance must start from its rest and move through an arc of about ten degrees before bringing the locking tooth into action.

TO TIGHTEN A COMMON PINION ON THE CENTRE ARBOR.

The most common way is to put a hair into the cannon and force it down upon the arbor, but this is objectionable from the fact that it sets the pinion just the width of the hair to one side. Another way is to twist the arbor lightly into a pair of cutting plyers, raising a thread or burr upon it. I could not recommend this mode as there is too much danger of bending the arbor in the operation. I generally roll the arbor between two files, letting the square part be to one side of them, of course. A very slight roll between two files will generally tighten the cannon, and there can be no danger of bending the arbor or setting the pinions to one side. '

TO TIGHTEN A RUBY PIN.

Set the ruby pin in asphaltum varnish. It will become hard in a few minutes, and be much firmer and better than gum shellac, as generally used.

CHAPTER IV.

ON MENDING WATCH TRAINS.

WHEN a wheel or a pinion is wanting in the train of a watch, it is usual to say the train is broken; and the act of supplying that wheel or pinion is generally termed mending the train. This, according to the old plan of working involved no small amount of labor, in the way of calculations, to get the proper size of the new piece. A person was under the necessity of being a good algebra scholar to do it. The recent, or I might say the American system— for European watchmakers still hold to their old ways— makes it much easier. A few simple tables have been gotten up by which any person who knows how to count and to measure may select the piece he wants in a few minutes.

TO DETERMINE THE REQUIRED DIAMETER OF A PINION.

For size of Pinion with	Measures on Wheel.	Character of Measure.
4 leaves, ...	2 teeth,......	Very full from out to out.
5 leaves, ...	3 teeth,......	Exactly from centre to centre.
6 leaves, ...	3 teeth,......	Full from centre to centre.
7 leaves, ...	3 teeth,......	Scant from out to out.
8 leaves, ...	4 teeth,......	Scant from centre to centre.
9 leaves, ...	4 teeth,......	Full from out to out.
10 leaves, ...	4 teeth,......	Exactly from out to out.
12 leaves	5 teeth,......	Exactly from centre to centre.
14 leaves, ...	6 teeth,......	Scant from centre to centre.
15 leaves.	6 teeth,......	Scant from out to out.
17 leaves, ...	7 teeth,......	Full from centre to centre.

TABLES OF NON-SECOND WATCH TRAINS.

Centre wheel. No. of teeth in wheel.	3d Wheel and Pinion. Teeth in wheel.	Leaves in pin.	4th Wheel and Pinion. Teeth in wheel.	Leaves in Pin.	Seconds in revolutions.	Scape Wheel and Pinion. Teeth in wheel.	Leaves in pin.	Beats per minute. No: of beats.	Character of trains.
66	63	6	63	6	31	7	6	283 scant	Trains for seven teeth in scape wheel.
66	64	6	63	6	31	7	6	287 full	
66	64	6	64	6	31	7	6	292 full	
72	66	6	58	6	27	7	6	298 scant	
66	63	6	62	6	31	7	6	278 full	
66	63	6	61	6	31	7	6	274 scant	
66	63	6	60	6	31	7	6	267 full	
63	60	6	56	6	34	9	6	294......	Trains for nine teeth in scape whee
66	60	6	54	6	33	9	6	297......	
63	60	6	57	6	34	9	6	299 full	
66	60	6	53	6	33	9	6	291 full	
63	60	6	55	6	34	9	6	289 scant	
60	60	6	52	6	33	9	6	286......	
63	60	6	54	6	34	9	6	283 full	
66	60	6	51	6	33	9	6	280 full	
63	60	6	53	6	34	9	6	278 full	
63	60	6	52	6	34	9	6	273......	
66	60	6	50	6	33	9	6	275......	
58	56	6	53	6	40	11	6	292 full	Trains for eleven teeth in scape wheel.
64	52	6	52	6	30	11	6	294 scant	
60	56	6	52	6	30	11	6	230 scant	
60	60	6	49	6	36	11	6	300 scant	
60	54	6	54	6	40	11	6	397......	
60	54	6	53	6	40	11	6	291 full	
62	54	6	51	6	39	11	6	290 scant	
58	54	6	54	6	41	11	6	287 full	
58	55	6	53	6	41	11	6	287......	
59	54	6	53	6	41	11	6	286 full	
60	54	6	52	6	40	11	6	286......	
61	55	6	51	6	39	11	6	286 scant	

TABLES OF NON-SECOND WATCH TRAINS.

(Continued.)

Centre wheel. No. of teeth in wheel.	3d Wheel and Pinion. Teeth in wheel.	Leaves in pin.	4th Wheel and Pinion. Teeth in wheel.	Leaves in Pin.	Seconds in revolutions.	Scape Wheel and Pinion. Teeth in wheel.	Leaves in pin.	Beats per minute. No. of beats.	Character of trains.
56	55	6	50	6	39	11	6	285 scant	
60	55	6	48	6	38	11	6	282 full	
62	54	6	52	6	41	11	6	281 scant	
63	54	6	51	6	40	11	6	281 full	
63	54	6	50	6	39	11	6	280 scant	
70	54	6	54	6	43	11	6	277 full	
70	60	6	48	6	36	11	6	293 full	
70	54	6	52	6	39	11	6	295 full	
60	54	6	50	6	38	11	6	289 scant	
63	48	6	56	6	43	11	6	287 full	
63	70	7	56	7	36	11	7	293 full	
80	70	7	48	7	36	11	6	293 full	
80	60	7	48	6	36	11	6	293 full	
80	70	6	48	7	36	11	6	293 full	
80	50	6	56	7	40	11	6	287 full	
80	63	6	50	7	38	11	6	289 scant	
80	80	8	64	8	36	11	8	293 full	
70	80	8	56	8	36	11	7	293 full	
70	80	8	48	8	36	11	6	293 full	
63	56	6	56	7	40	11	6	287 full	
63	64	6	56	8	40	11	6	287 full	
84	48	8	56	6	40	11	6	287 full	
84	56	8	56	7	40	11	6	287 full	
84	64	8	56	8	40	11	6	287 full	
63	63	6	50	7	38	11	6	289 scant	
63	72	6	50	8	38	11	6	289 scant	
84	54	8	50	6	38	11	6	289 scant	
84	63	8	50	7	38	11	6	289 scant	
60	72	8	50	8	38	11	6	289 scant	
70	80	6	48	8	36	11	6	293 full	

Trains for eleven teeth in scape wheel.

TABLES OF NON-SECOND WATCH TRAINS.

(Continued.)

Centre wheel. No. of teeth in wheel.	3d Wheel and Pinion. Teeth in wheel.	Leaves in pin.	4th Wheel and Pinion. Teeth in wheel.	Leaves in Pin.	Seconds in revolutions.	Scape Wheel and Pinion. Teeth in wheel.	Leaves in pin.	Beats per minute. No. of beats.	Character of trains.
70	80	7	48	8	36	11	6	293 full	Trains for eleven teeth in scape wheel.
80	80	7	56	8	36	11	7	293 full	
80	60	8	48	6	36	11	6	293 full	
80	70	8	48	7	36	11	6	293 full	
80	70	8	56	7	36	11	7	293 full	
52	52	6	51	6	48	13	6	277 scant	Trains for thirteen teeth in scape wheels.
57	51	6	48	6	44	13	6	280 scant	
56	51	6	49	6	45	13	6	281 scant	
54	52	6	50	6	46	13	6	282 scant	
50	51	6	50	6	45	13	6	284 full	
54	43	6	50	6	45	13	6	287 full	
54	52	6	51	6	46	13	6	287 full	
57	53	6	48	6	43	13	6	291 scant	
56	54	6	48	6	44	13	6	291 full	
56	53	6	49	6	44	13	6	292 scant	
54	53	6	51	6	45	13	6	293 scant	
60	51	6	48	6	42	13	6	294 full	
59	51	6	49	6	43	13	6	296 scant	
56	53	6	50	6	44	13	6	298 scant	
54	53	6	52	6	45	13	6	298 full	
53	52	6	50	6	46	13	6	276 full	
52	52	6	52	6	46	13	6	293 scant	
55	51	6	51	6	46	13	6	287	
56	50	6	51	6	46	13	6	286 full	
56	52	6	48	6	44	13	6	280 full	
56	52	6	50	6	44	13	6	292 full	
60	48	6	48	6	45	13	6	277 full	
60	50	6	48	6	43	13	6	289 scant	
60	54	6	60	8	53	13	6	292 full	
60	58	7	56	7	51	13	6	287 full	

TABLES OF NON-SECOND WATCH TRAINS.

(Continued)

Centre wheel.	3d Wheel and Pinion.		4th Wheel and Pinion.			Scape Wheel and Pinion.		Beats per minute.	Character of trains.
No. of teeth in wheel.	Teeth in wheel.	Leaves in pin.	Teeth in wheel.	Leaves in Pin.	Seconds in revolutions	Teeth in wheel.	Leaves in pin.	No. of beats.	
60	60	8	54	6	44	13	6	300	Trains for thirteen teeth in scape wheel.
62	56	7	56	7	47	13	6	396 full	
63	52	7	51	6	60	13	6	285	
63	60	7	60	7	60	13	6	290	
64	60	7	60	7	60	13	6	285	
72	70	8	68	8	60	13	6	280	
74	68	8	68	8	60	13	6	286 full	
48	45	6	56	6	50	15	6	288	Trains for fifteen teeth in scape wheel.
48	45	6	57	6	62	15	6	288	
48	45	6	58	6	62	15	6	300	
48	45	6	59	6	60	15	6	291 scant	
58	48	6	46	6	50	15	6	290	
54	50	6	48	6	48	15	6	286	
56	48	6	46	6	50	15	6	289 scant	
63	56	7	56	7	50	15	7	288	
60	56	8	58	7	50	15	6	288	
62	60	8	60	8	50	15	6	288	
72	64	8	50	8	50	15	6	288	
72	64	8	56	8	50	15	7	288	
72	64	8	64	8	50	15	8	288	
52	50	6	48	6	50	15	6	288	
54	48	6	48	6	50	15	6	288	
72	64	8	48	8	50	15	6	288	
72	80	8	64	10	50	15	8	288	
72	80	8	56	10	50	15	7	288	
72	80	8	48	10	50	15	6	288	
63	80	7	64	10	50	15	8	288	
63	80	7	56	10	50	15	7	288	
63	80	7	48	10	50	15	6	288	
..	..	.	..	..	..	.	.	....	

TABLES OF NON-SECOND WATCH TRAINS.
(Continued.)

Centre wheel.	3d Wheel and Pinion.		4th Wheel and Pinion.			Scape Wheel and Pinion.		Beats per minute.
No. of teeth in wheel.	Teeth in wheel.	Leaves in pin.	Teeth in wheel.	Leaves in Pin.	Seconds in revolutions	Teeth in wheel.	Leaves in pin.	No. of beats.
72	64	8	56	8	50	17	8	286 scant.
64	64	8	64	8	50	17	8	290 full.
48	48	6	45	6	53	17	6	272 ......
48	48	6	46	6	53	17	6	278 ......
64	80	8	48	10	53	17	6	299 full.
54	48	6	44	6	50	17	6	299 full.
51	48	6	45	6	53	17	6	295 full.
54	48	6	43	6	50	17	6	292 full.
48	48	6	48	6	53	17	6	290 full.
51	48	6	45	6	53	17	6	289 ......
54	48	6	42	6	53	17	6	286 scant.
48	48	6	47	6	53	17	6	284 full.
51	48	6	44	6	53	17	6	283 scant.
64	64	8	60	8	53	17	8	289 scant.
56	56	7	56	7	53	17	7	290 full.
63	56	7	49	7	53	17	7	286 scant.
64	56	8	48	7	53	17	6	290 full.
80	80	10	64	10	53	17	8	290 full.
80	64	10	64	8	53	17	8	290 full.
80	64	10	56	8	53	17	7	290 full.
80	64	10	48	8	53	17	6	290 full.
80	56	10	56	7	53	17	7	290 full.
80	56	10	48	7	53	17	6	290 full.
64	80	8	64	10	53	17	8	290 full.
64	80	8	56	10	53	17	7	290 full.

Trains for seventeen teeth in scape wheel.

TABLES OF FOURTH WHEEL SECOND WATCH TRAINS.

Centre wheel. No. of teeth in wheel.	3d Wheel and Pinion. Teeth in wheel.	Leaves in pin.	4th Wheel and Pinion. Teeth in wheel.	Leaves in Pin.	Seconds in revolutions	Scape Wheel and Pinion. Teeth in wheel.	Leaves in pin.	Beats per minute. No. of beats.	
48	45	6	76	6	60	11	6	279 scant.	
48	45	6	74	6	60	11	6	271 full.	
48	45	6	71	6	60	11	6	260 full.	
56	60	7	74	8	60	11	6	271 full.	
48	75	6	78	6	60	11	6	286	
60	79	7	74	7	60	11	6	271 full.	
60	79	7	76	7	60	11	6	279 scant.	
60	79	7	78	7	60	11	6	286	
45	56	6	74	7	60	11	6	271 full.	
45	56	6	76	7	60	11	6	279 scant.	
45	56	6	78	7	60	11	6	286	
64	60	8	74	8	60	11	6	271 full.	Fourth wheel seconds with eleven teeth in scape wheel.
64	60	8	76	8	60	11	6	279 scant.	
64	60	8	78	8	60	11	6	286	
60	56	8	74	7	60	11	6	271 full.	
60	56	8	78	7	60	11	6	286	
60	78	8	74	6	60	11	6	271 full.	
48	78	8	78	6	60	11	6	286	
48	60	6	74	8	60	11	6	271 full.	
48	60	6	78	8	60	11	6	286	
60	56	8	76	7	60	11	6	279 scant.	
64	60	8	66	8	60	13	6	286	Fourth wheel seconds with 13 teeth in scape wheel.
64	60	8	67	8	60	13	6	290 full.	
56	75	7	68	10	60	13	6	295 scant.	
45	56	6	66	7	60	13	.6	286	
60	49	7	66	7	60	13	6	286	

TABLES OF FOURTH WHEEL SECOND WATCH TRAINS.

(Continued.)

Centre wheel. No. of teeth in wheel.	3d Wheel and Pinion. Teeth in wheel.	Leaves in pin.	4th Wheel and Pinion. Teeth in wheel.	Leaves in Pin.	Seconds in revolutions.	Scape Wheel and Pinion. Teeth in wheel.	Leaves in pin.	Beats per minute. No. of beats.	
60	49	7	77	7	60	13	7	286	Fourth wheel seconds with thirteen teeth in scape wheel
64	60	8	69	8	60	13	6	299	
64	60	8	68	8	60	13	6	295 scant.	
60	49	7	67	7	60	13	6	290 full.	
48	45	6	66	6	60	13	6	286	
48	45	6	67	6	60	13	6	290 full.	
48	45	6	68	6	60	13	6	264 scant.	
48	45	6	69	6	60	13	6	299	
60	56	8	66	7	60	13	6	286	
80	60	10	66	8	60	13	6	286	
64	75	8	66	10	60	13	6	286	
48	60	6	66	8	60	13	6	286	
48	75	6	66	10	60	13	6	286	
64	45	8	60	6	60	15	6	300	Fourth wheel seconds with fifteen teeth in scape wheel.
64	60	8	60	8	60	15	6	300	
64	64	8	70	10	60	15	7	300	
64	60	8	70	8	60	15	7	300	
60	56	8	60	7	60	15	6	300	
48	60	6	60	8	60	15	6	300	
60	70	7	70	7	60	15	7	300	
60	49	7	60	7	60	15	6	300	
48	49	6	60	6	60	15	6	300	
80	45	10	70	8	60	15	7	300	
75	60	10	60	8	60	15	6	300	
64	75	7	60	10	60	15	6	300	

TABLES OF FOURTH WHEEL SECOND WATCH TRAINS.
(Continued.)

Centre wheel.	3d Wheel and Pinion.		4th Wheel and Pinion.			Scape Wheel and Pinion.		Beats per minute.	
No. of teeth in wheel.	Teeth in wheel.	Leaves in pin.	Teeth in wheel.	Leaves in Pin.	Seconds in revolutions	Teeth in wheel.	Leaves in pin.	No. of beats.	
56	75	7	70	10	60	15	7	300	Fourth wheel seconds with fifteen teeth in scape wheel.
56	75	8	60	10	60	15	6	300	
64	75	8	54	8	60	15	6	270	
60	60	8	54	7	60	15	6	270	
64	56	8	54	6	60	15	6	270	
48	45	6	54	8	60	15	6	270	
60	60	7	63	7	60	15	7	270	
60	56	8	48	7	60	15	6	240	
60	49	7	54	7	60	15	6	270	
48	49	6	54	6	60	15	6	270	
64	45	8	48	8	60	15	6	240	
60	60	8	48	7	60	15	6	240	
48	50	6	48	8	60	15	6	240	
64	60	8	48	6	60	15	6	240	
60	45	7	56	7	60	15	7	240	
60	49	7	48	7	60	15	6	240	
48	45	6	48	6	60	15	6	240	
60	56	8	51	7	60	17	6	289	Fourth wheel seconds with seventeen teeth in scape wheel.
64	60	8	50	8	60	17	6	283 full.	
64	60	8	51	8	60	17	6	289	
75	56	10	68	7	60	17	8	289	
80	60	10	50	8	60	17	6	283 full.	
75	64	10	50	8	60	17	6	283 full.	
75	68	10	68	8	60	17	8	289	
80	75	10	68	10	60	17	8	289	

TABLES OF FOURTH WHEEL SECOND WATCH TRAINS.

(Continued.)

Centre wheel.	3d Wheel and Pinion.		4th Wheel and Pinion.			Scape Whee and Pinion		Beats per minute.	
No. of teeth in wheel.	Teeth in wheel.	Leaves in pin.	Teeth in wheel.	Leaves in Pin.	Seconds in revolutions	Teeth in wheel.	Leaves in pin.	No. of beats.	Third wheel and patent second trains.
60	72	6	60	12	60		6	300	
60	60	6	60	10	60		6	300	
60	48	6	60	8	60		6	300	
48	60	6	60	8	60		6	300	
48	60	6	54	8	60		6	270	
60	72	6	54	12	60		6	270	
48	60	6	48	8	60		6	240	
60	60	6	54	10	60		6	270	
60	72	6	48	12	60		6	240	
48	60	6	48	10	60		6	240	

American Watch.

64	60	8	64	8	60	15	7	300 . . . . . .

Trial Watch.

80	75	10	80	10	60	15	8	300 . . . . . .

CHAPTER V.

ON TEMPERING.

No part of his trade gives the self-instructed watchmaker more trouble than the acquirement of an ability to temper, as they should be, his various tools and pieces of machinery; in fact a whole life devoted to experiments and study touching this department, would not be likely to attain the desired end. And yet all the processes employed are so amazingly simple as to lead one to wonder, when he understands them, *why* he did not know all about them before.

TO TEMPER BRASS, OR TO DRAW ITS TEMPER.

Brass is rendered hard by hammering or rolling, therefore when you make a thing of brass, necessary to be in temper, you must prepare the material before shaping the article. Temper may be drawn from brass by heating it to a cherry red, and then simply plunging it into water the same as though you were going to temper steel.

TO TEMPER DRILLS.

Select none but the finest and best steel for your drills. In making them never heat higher than a cherry red, and always hammer till nearly cold. Do all your hammering in one way, for if, after you have flattened your piece out, you attempt to hammer it back to a square or a round you spoil it. When your drill is in proper shape heat it to a cherry red, and thrust it into a piece of resin, or into quicksilver.

Some use a solution of cyanuret potassa and rain-water for tempering their drills, but for my part I have always found the resin or quicksilver to work best.

TO TEMPER GRAVERS.

Gravers and other instruments larger than drills, may be tempered in quicksilver as above; or you may use lead instead of quicksilver. Cut down into the lead, say half an inch ; then, having heated your instrument to a light cherry red, press it firmly into the cut. The lead will melt around it, and an excellent temper will be imparted.

TO TEMPER CASE SPRINGS.

Having fitted the spring into the case according to your liking, temper it hard by heating and plunging into water. Next polish the small end so that you may be able to see when the color changes ; lay it on a piece of copper or brass plate, and hold the plate over your lamp, with the blaze directly under the largest part of the spring. Watch the polished part of the steel closely, and when you see it turn blue remove the plate from the lamp, letting all cool gradually together. When cool enough to handle polish the end of the spring again, place it on the plate and hold over the lamp as before. The third bluing of the polished end will leave the spring in proper temper. Any steel article to which you desire to give a spring temper may be treated in the same way.

Another process said to be good—I have never tried it— is to temper the spring as in the first instance; then put it into a small iron ladle, cover it with linseed oil and hold over a lamp till the oil takes fire. Remove the ladle, but let the oil continue to burn until nearly all consumed, when blow out, re-cover with oil and hold over the lamp as before. The third burning out of the oil will leave the spring in the right temper.

TO TEMPER CLICKS, RATCHETS, ETC.

Clicks, ratchets or other steel articles requiring a similar degree of hardness should be tempered in mercurial ointment. The process consists in simply heating to a cherry red and plunging into the ointment. No other mode will combine toughness and hardness to such an extent.

TO DRAW THE TEMPER FROM DELICATE STEEL PIECES WITHOUT SPRINGING THEM.

Place the articles from which you desire to draw the temper into a common iron clock key. Fill around it with brass or iron filings, and then plug up the open end with a steel, iron or brass plug, made to fit closely. Take the handle of the key with your plyers and hold its pipe into the blaze of a lamp till near hot, then let it cool gradually. When sufficiently cold to handle, remove the plug, and you will find the article with its temper fully drawn, but in all other respects just as it was before.

You will understand the reason for having the article thus plugged up while passing it through the heating and cooling process, when I tell you that springing always results from the action of changeable currents of atmosphere. The temper may be drawn from cylinders, staffs, pinions, or any other delicate pieces by this mode with perfect safety.

TO TEMPER STAFFS, CYLINDERS OR PINIONS, WITHOUT SPRINGING THEM.

Prepare the articles as in preceding process, using a steel plug. Having heated the key-pipe to a cherry red, plunge it into water; then polish the end of your steel plug, place the key upon a plate of brass or copper, and hold it over your lamp with the blaze immediately under the pipe till the polished part becomes blue. Let cool gradually, then polish again. Blue and cool a second time, and the work will be done.

TO DRAW THE TEMPER FROM PART OF A SMALL STEEL ARTICLE.

Hold the part from which you wish to draw the temper, with a pair of tweezers, and with your blow-pipe direct the flame upon them—not the article—till sufficient heat is communicated to the article to produce the desired effect.

TO BLUE SCREWS EVENLY.

Take an old watch barrel and drill as many holes into the head of it as you desire to blue screws at a time. Fill it about one-fourth full of brass or iron filings, put in the head, and then fit a wire, long enough to bend over for a handle, into the arbor holes—head of the barrel upwards. Brighten the heads of your screws, set them, point downwards, into the holes already drilled, and expose the bottom of the barrel to your lamp till the screws assume the color you wish

TO REMOVE BLUING FROM STEEL.

Immerse in a pickle composed of equal parts muriatic acid and elixir vitriol. Rinse in pure water and dry in tissue paper.

TO CASE-HARDEN IRON.

Heat to a bright red in a crucible or ladle; pour in enough powdered cyanid of potash to cover it; let remain five or six seconds, and then turn out into rain water. The piece treated in this way will polish up equal to steel, and be almost quite as hard.

—o○o○oo—

CHAPTER VI.

ON MILLS, BROACHES, FILES AND BURNISHERS.

YOUR diamond mills, diamond broaches and diamond files you can generally buy ready made to suit, though instances may occur in which you will require them of a peculiar size and shape, not to be had of the dealers. It is, therefore, best to know how to prepare them. I make all my own for two reasons—they are better than those I can buy, and they do not cost me anything like as much.

To make these articles diamond dust is necessary. This you can buy in most of the large cities ready prepared. It is not a costly article; one dollar's worth will last you a long time.

TO MAKE A DIAMOND MILL.

Make a plain brass wheel about two inches in diameter, and arrange it to work to your foot-lathe. Place it flat on some solid substance, and having oiled its face, sprinkle it thinly with coarse diamond dust. With a smoothe hammer then tap it lightly till the diamond dust is thoroughly driven into the brass. The brass will bur around it and hold it securely in place. We use the oil to prevent the dust from bounding off while undergoing the process of hammering.

A mill prepared in this way will last for years. I have one now in my shop, upon which I have ground watch, spectacle and breastpin glasses for five years, and yet it appears as sharp, and cuts as well as it did at first. As the wheel wears off the diamond grains seem to sink into the brass from the effect of the grinding.

TO MAKE DIAMOND BROACHES.

Make your broaches of brass the size and shape you desire; then, having oiled them slightly, roll their points into fine diamond dust till entirely covered. Hold them then on the face of your anvil and tap with a light hammer till the grains disappear in the brass. Great caution will be necessary in this operation. Do not tap heavy enough to flatten the broach. Very light blows are all that will be required; the grains will be driven in much sooner than one would imagine.

Some roll the broach between two smoothe pieces of steel to imbed the diamond dust. It is a very good way, but somewhat more wasteful of the dust.

Broaches made on this plan are used for dressing out jewels.

TO MAKE POLISHING BROACHES.

These are usually made of ivory, and used with diamond dust, loose, instead of having been driven in. You oil the broach lightly, dip it into the finest diamond dust and proceed to work it into the jewel the same as you do the brass broach. Unfortunately too many watchmakers fail to attach sufficient importance to the polishing broach. The sluggish motion of watches now-a-days, is more often attributable to rough jewels than to any other cause.

TO MAKE DIAMOND FILES.

Shape your file of brass, and charge with diamond dust, as in case of the mill. Grade the dust in accordance with the coarse or fine character of the file desired.

TO MAKE PIVOT FILES.

Dress up a piece of wood file fashion, about an inch

broad, and glue a piece of fine emery paper upon it. Shape your file then, as you wish it, of the best cast-steel, and before tempering pass your emery paper heavily across it several times, diagonally. Temper by heating to a cherry red, and, plunging into linseed oil. Old worn pivot files may be dressed over and made new by this process. At first thought one would be led to regard them too slightly cut to work well, but not so. They dress a pivot more rapidly than any other file.

TO MAKE BURNISHERS.

Proceed the same as in making pivot files, with the exception that you are to use fine flour of emery on a slip of oiled brass or copper, instead of the emery paper. Burnishers which have become too smoothe may be improved vastly with the flour of emery as above without drawing the temper.

TO PREPARE A BURNISHER FOR POLISHING.

Melt a little beeswax on the face of your burnisher. Its effect then, on brass or other finer metals, will be equal to the best buff. A small burnisher prepared in this way is the very thing with which to polish up watch wheels. Rest them on a piece of pith while polishing.

CHAPTER VII.

ON CLEANING AND REPAIRING CLOCKS.

THE clocks now generally in use among our people are so simple in their construction, and the processes employed to keep them in order are so few and plain, that a lengthy treatise on the subject as indicated by the above heading, could hardly be profitable. Almost any person endowed with common sense and a taste for working at light machinery, may, with a little practice, clean and repair clocks successfully.

With all its simplicity, however, there are many persons following the business of cleaning and repairing clocks who do not give satisfaction; or, in other words, who do not seem to possess all the necessary requisites. As an illustration—a man will come to your house, perhaps, take down your clock, clean it properly, repair it all right, put it up as it should be, and then—spoil the job by oiling all the pivots and probably the pinions. The requisite lacking in this case is good common sense. If he had possessed this he would have seen that to oil the pivots or pinions would be to cause their accumulation of dust; that this dust mixing with the oil, must increase the friction by causing the parts to grind together, to say nothing of a gum sure to result —either of which, without the other, could not do otherwise than stop the machine sooner or later,

We often hear persons complaining of their clocks stopping in cold weather—in nine cases out of ten the cause may be attributed to this very injudicious use of oil. A gum has formed on the pivots or pinions, or both, which stiffens under the influence of the cold, and, of course, stops the movement. But this is not the only bad result. A clock grinding along in consequence of having been improperly oiled, will wear out in less than half the time that it would under other circumstances. The reason in this must be apparent to all—each pivot or each pinion leaf has been converted, as it were, into a grindstone.

I am sorry to say that a large per cent. of the professed clock-tinkerers straggling over our country do work on the plan just named. They are generally men who are too lazy to earn an honest living by hard labor, and too dull to do it in any other way. If a man is disposed to work at clocks, and possesses the requirements that will enable him to do it well, a necessity for much "tramping" will never spring up. A community can easily be found that will give him a permanent business. And unless the person applying for a "job" is known, or can furnish satisfactory evidence that he understands his business, and is honest enough to do well what he understands, my advice is to keep him and the clock as far apart as possible. Better ten to one that the owner go to work and put it in repair himself; for certain it is that he will not willfully injure his own property.

Under the impression then that this book may possibly fall into the hands of some who, in consequence of not being convenient to the establishment of a regular watch or clock repairer, would like to keep their own clocks in order, I shall proceed to give a few simple directions, which, if followed, will enable them to do so without trouble.

TO CLEAN A CLOCK.

Take the movement of the clock "to pieces." Brush the wheels and pinions thoroughly with a stiff, coarse brush; also the plates into which the trains work. Clean the pivots well by turning in a piece of cotton cloth held tightly between your thumb and finger. The pivot holes in the plates are generally cleansed by turning a piece of wood into them, but I have always found a strip of cloth or a soft cord drawn tightly through them to act the best. If you use two cords, the first one slightly oiled, and the next dry to clean the oil out, all the better. Do not use salt or acid to clean your clock—it can do no good, but may do a great deal of harm. Boiling the movement in water, as some practice, is also foolishness.

TO BUSH.

The holes through which the great arbors, or winding axles work, are the only ones that usually require bushing. When they have become too much worn the great wheel on the axle before named strikes too deeply into the pinions above it, and stops the clock. To remedy this bushing is necessary, of course. The most common way of doing it is to drive a steel point or punch into the plate just above the axle hole, thus forcing the brass downward until the hole is reduced to its original size. Another mode is to solder a piece of brass upon the plate in such a position as to hold the axle down to its proper place. If you simply wish your clock to run, and have no ambition to produce a bush that will look workmanlike, about as good a way as any is to fit a piece of hard wood between the post which comes through the top of the plate and the axle. Make it long enough to hold the axle to its proper place, and so that the axle will run on the end of the grain. Cut notches where the pivots come through, and secure by wrapping around it and the plate a piece of small wire, or a thread. There is no post coming through above the axle on the striking side, but this will rarely require pushing. I have known clocks to run well on this kind of bushing, botchified as it may appear, for ten years.

TO REMEDY WORN PINIONS.

Turn the leaves or rollers so the worn places upon them will be towards the arbor or shaft, and fasten them in that position. If they are " rolling pinions," and you cannot secure them otherwise, you had better do it with a little soft solder.

TO OIL PROPERLY.

Oil only, and very lightly, the pallets of the verge, the steel pin upon which the verge works, and the point where the loop of the verge wire works over the pendulum wire. Use none but the best watch oil. Though you might be working constantly at the clock-repairing business, a bottle costing you but twenty-five cents, would last you two years at least. You can buy it at any watch-furnishing establishment.

TO MAKE THE CLOCK STRIKE CORRECTLY.

If not very cautious in putting up your clock you will get some of the striking-train wheels in wrong, and thus produce a derangement in the striking. If this should happen, prize the plates apart on the striking side, slip the pivots of the upper wheels out, and having disconnected them from the train, turn them part around and put them back. If still not right, repeat the experiment. A few efforts at most will get them to working properly.

A DEFECT TO LOOK AFTER.

Always examine the pendulum wire at the point where the loop of the verge wire works over it. You will generally find a small notch, or at least a rough place, worn there. Dress it out perfectly smooth, or your clock will not be likely to work well. Small as this defect may seem, it stops a large number of clocks.

CHAPTER VIII.

ON REFINING AND COMPOUNDING METALS.

ALTHOUGH it is not expected that the watchmaker and jeweler will be called upon to do a heavy business in the way of refining metals, yet it is proper for him to know something of the *modus operandi*, for cases may occur in which it will be necessary for him to separate the members of a compound, or to have a metal which he can rely upon as being pure. I shall, therefore, lay before him a few simple recipes. They are not exactly the processes employed when refining is done on a large scale, but they are perfectly reliable, and will answer his purpose ; in fact they are the only ones he could make use of without extensive and expensive preparations.

A thorough knowledge of the formula by which metals are compounded is of the utmost importance.

TO REFINE GOLD.

If you desire to refine your gold from the baser metals, swedge or roll it out very thin, then cut into narrow strips and curl up so as to prevent its lying flatly. Drop the pieces thus prepared into a vessel containing good nitric acid, in the proportion of acid two ounces, and pure rain water half an ounce. Suffer to remain until thoroughly dissolved, which will be the case in from half an hour to one hour. Then pour off the liquid carefully and you will find the gold in the form of a yellow powder lying at the bottom of the vessel. Wash this with pure water till it ceases to have an acid taste, after which you may melt and cast into any form you choose. Gold treated in this way may be relied on as perfectly pure.

In melting gold use none other than a charcoal fire, and during the process sprinkle saltpetre and potash into the crucible occasionally. Do not attempt to melt with stone coal, as it renders the metal brittle and otherwise imperfect.

TO REFINE SILVER.

Dissolve in nitric acid as in the case of the gold. When

the silver has entirely disappeared, add to the two-and-a-half ounces of solution nearly one quart of pure rain water. Sink, then, a sheet of clean copper into it—the silver will collect rapidly upon the copper, and you can scrape it off and melt into bulk at pleasure.

In the event you were refining gold in accordance with the foregoing formula, and the impurity was silver, the only steps necessary to save the latter would be to add the above named proportion of water to the solution poured from the gold, and then to proceed with your copper plate as just directed.

TO REFINE COPPER.

This process differs from the one employed to refine silver in no respects save the plate to be immersed—you use an iron instead of a copper plate to collect the metal.

If the impurities of gold refined were both silver and copper, you might, after saving the silver as above directed, sink your iron plate into the solution yet remaining, and take out the copper. The parts of alloyed gold may be separated by these processes. and leave each in a perfectly pure state.

TO MAKE COIN GOLD.

The gold of American and English coin is twenty-two carat fine. Copper alone usually forms the alloy, though a portion of silver is sometimes added. To make coin gold, you melt together with saltpetre and sal-ammoniac, the two metals in the proportion of twenty-two grains pure gold and two grains pure copper. When silver forms a part of the alloy it is usually about one-third silver to two-thirds copper. The latest American coin is of that alloy.

TO MAKE EIGHTEEN CARAT GOLD

To make the eighteen carat gold, generally in use, melt together as above, eighteen grains pure gold, four grains pure copper and two grains pure silver. In cases where you find it necessary to use gold coin, weigh out in the proportion of nineteen-and-a-half grains gold, three grains copper and one-and-a-half grains silver.

TO MAKE SIXTEEN CARAT GOLD

Compound sixteen grains pure gold with five-and-a-half grains pure copper and two-and-a-half grains pure silver. Or, if gold coin is used, seventeen grains gold, five grains copper and two grains silver.

TO MAKE TWELVE CARAT GOLD.

Melt together, in the usual way, twenty-five grains gold —if coin—thirteen-and-a-half grains copper, and seven-and-a-third grains silver. This is a very good gold for rings, &c.—stands acids almost equal to the higher grades, and looks fully as well. Of course it is deficient in weight.

TO MAKE FOUR CARAT GOLD.

Four carat gold is used to a considerable extent for cheap rings, pin-tongues and the like. It is a very nice metal, wears well, does not black the finger, and presents somewhat the appearance of Guinea gold. You make it by melting together eighteen parts copper, four parts gold, and two parts silver

TO MAKE GREEN GOLD.

Melt together nineteen grains pure gold and five grains pure silver. The metal thus prepared has a beautiful green shade. Some years ago it was used pretty extensively by jewelers in the formation of leaves but we do not meet with it so often now.

TO MAKE BEST COUNTERFEIT GOLD

Fuse together with saltpetre, sal-ammoniac and powdered charcoal, four parts platina, two-and-a-half parts pure copper, one part pure zinc, two parts block tin and one-and-a-half parts pure lead.

Another good recipe calls for two parts platina, one part silver and three parts copper.

A metal compounded in accordance with either formula, as exhibited above, will so nearly resemble gold as to almost defy detection without a resort to thorough tests. The platina requires a high temperature to melt, but nothing could be substituted that would act so well, as it adds to the ring of the metal, and to a great extent fortifies it against the action of acids.

If at any time you should find your metal too hard or brittle for practical use re-melt it with sal-ammoniac. It may in some cases be necessary to repeat this operation several times, but it will bo sure to produce the desired effect eventually.

TO MAKE BEST OREIDE GOLD.

Oreide gold is figuring no little at this time in the way of cheap jewelry. The best article is made by compounding four parts pure copper, one-and-three-fourth parts pure zinc, one-fourth part magnesia, one-tenth part sal-ammoniac, one-twelfth part quick-lime and one part cream tartar. Melt the copper first, then add as rapidly as possible the other articles in the order named.

TO MAKE ALLOYED SILVER.

Copper is the only less precious metal that alloys well with silver. Its addition is a decided improvement on the original, rendering it harder, finer in appearance and more sonorous; and it is astonishing to note the quantity that may be added without otherwise changing the first appearance of the metal. An alloy of silver and copper in the proportion of four-fifths silver to one of copper, is fully as white as the silver would be entirely pure. When the proportion of copper rises above one-fifth, it begins to have an influence in the color. American coin silver is one-tenth copper.

The baser white metals cannot be alloyed with silver to any great extent, owing to the fact that they impart to the compound too great a degree of brittleness. A small proportion of block tin virtually converts it into bell metal.

The following is, perhaps, the best known composition for a cheap silver : Pure silver, say one ounce; copper, one-sixth of an ounce; brass, two-thirds of an ounce; bismuth, one-third of an ounce; clean salt, two-thirds of an ounce; white arsenic, one-third of an ounce; and potash, one-third of an ounce. Melt the silver, copper and brass first, then add the other articles in the order named. Sprinkle a very little borax into the crucible while melting—too much will have a tendency to render the metal unmalleable.

TO MAKE BEST COUNTERFEIT SILVER.

Combine by fusion one part pure copper, twenty-four parts block tin, one-and-a-half parts pure antimony, one-fourth part pure bismuth and two parts clear glass. The glass may be omitted save in cases where it is an object to have the metal sonorous.

TO MAKE GERMAN SILVER.

The best German silver may be made by melting together twenty-five parts copper, fifteen parts zinc and ten parts nickel.

TO MAKE GOLD SOLDER.

Melt together in a charcoal fire twenty-four grains gold coin, nine grains pure silver, six grains copper and three grains good brass. This makes a solder for gold ranging from twelve to sixteen carats in fineness. Where a finer grade is to be worked, the solder may be made to correspond by increasing the proportion of gold in its composition. A darker solder may be made, if desired, by lessening the proportion of silver, and increasing that of the copper in a corresponding degree.

TO MAKE SILVER SOLDER.

The usual method is to combine two parts of silver with one of brass. For my use I generally make the proportion of brass a little larger than one-third. In the course of his work the jeweler invariably throws aside quite a number of cheap pin-tongues as being too soft and too easily bent to be serviceable. Of these I often make my solder, combining them with silver in equal proportion. It seems to work better and more freely than any other I can prepare.

TO MAKE BRASS OR COPPER SOLDER.

Compound in the usual way two parts of brass with one of zinc. Such is the granulated solder sold in the shops under the name of *spelter*.

TO MAKE SOFT SOLDER.

The soft solder used by jewelers is generally a composition of two parts tin and one part lead. A solder composed of two parts bismuth, one part tin and one part lead, flows at a much lower temperature than the above; but it is not so strong.

———◆◇◆———

CHAPTER IX.

ON SOLDERING.

The first thing to be sure of in making preparations for soldering, is that the compound to be used in uniting the parts is easier of fusion than the parts themselves. Let this be otherwise and the attempt must certainly result in failure. The next thing to look after is the uniformity in the color of the solder and the metal to be soldered; and where such a thing is of importance, the uniformity in point of hardness. To have the color the same is often a matter of no little moment, especially in the case of rings, where the joint would otherwise be made to show. This last, though not least thing in point of consequence, is to see that the surfaces to be joined are perfectly bright and clean. Without this last-named precaution it is impossible to do good work.

TO HARD SOLDER GOLD, SILVER, COPPER, BRASS, IRON, STEEL OR PLATINA.

The solders to be used for gold, silver, copper and brass are given in the preceding chapter. You commence operations by reducing your solder to small particles and mixing it with powdered sal-ammoniac and powdered borax in equal parts, moistened to make it hold together. Having fitted up the joint to be soldered, you secure the article upon a piece of soft charcoal, lay your soldering mixture immediately over the joint, and then with your blow pipe turn the flame of your lamp upon it until fusion takes place. The job is then done and ready to be cooled and dressed up.

Iron is usually soldered with copper or brass in accordance with the above process. The best solder for steel is pure gold or pure silver, though gold or silver solders are often used successfully.

Platina can only be soldered well with gold; and the expense of it, therefore, contributes to the hinderance of a general use of platina vessels, even for chemical purposes, where they are of so much importance.

TO MAKE SOLDERING FLUIDS.

Clip into one ounce of best muriatic acid as much clear sheet zinc as it will dissolve; then add fifteen or twenty grains of sal-ammoniac and half an ounce of pure rain water.

The above fluid is not suitable for iron or steel, on account of the corrosive character of the acid. A soldering fluid may be made for those metals by dissolving chloride of zinc in alcohol. It does not run the solder quite so freely as does the first-named fluid, though it answers a very good purpose. These fluids are only used in soft soldering

TO SOFT SOLDER ARTICLES.

Moisten the parts to be united with soldering fluid; then, having joined them together, lay a small piece of solder upon the joint and hold over your lamp, or direct the blaze upon it with your blow-pipe until fusion is apparent. Withdraw then from the blaze immediately, as too much heat will render the solder brittle and unsatisfactory. When the parts to be joined can be made to spring or press against each other, it is best to place a thin piece of solder between them before exposing to the lamp.

· Where two smooth surfaces are to be soldered one upon the other, you may make an excellent job by moistening them with the fluid, and then, having placed a sheet of tin foil between them, holding them pressed firmly together over your lamp till the foil melts. If the surfaces fit nicely a joint may be made in this way so close as to be almost imperceptibla. The brightest looking lead which comes as a lining to tin boxes works better in the same way than tin foil.

TO CLEANSE GOLD TARNISHED IN SOLDERING.

The old English mode was to expose all parts of the article to a uniform heat, allow it to cool and then boil until bright in urine and sal-ammoniac. It is now usually cleansed with diluted sulphuric acid. The pickle is made in about the proportion of one-eighth of an ounce acid to one ounce rain water.

TO CLEANSE SILVER TARNISHED IN SOLDERING.

Some expose to a uniform heat, as in the case of gold, and then boil in strong alum water. Others immerse for a considerable length of time in a liquid made of half an ounce of cyanuret potassa to one pint rain water, and then brush off with prepared chalk.

———ⴰⴰ⚬⚬ⴰⴰ———

CHAPTER X.

ON PLATING.

To plate, according to the original meaning, was to solder a thin layer of gold or silver upon a baser metal, and then roll out the two together. In later days a broader meaning has been given to the word, so that any method of laying a finer metal upon a coarser is known as plating. There are now various modes of doing this, all of which are more or less interesting and useful to the watchmaker and jeweler.

TO MAKE GOLD SOLUTION FOR ELECTRO-PLATING.

Dissolve five pennyweights gold coin, five grains pure copper and four grains pure silver in three ounces nitro-muriatic acid; which is simply two parts muriatic acid and one part nitric acid. The silver will not be taken into solution as are the other two metals, but will gather at the bottom of the vessel. Add one ounce pulverized sulphate of iron, half an ounce pulverized borax, twenty-five grains pure table salt, and one quart hot rain water. Upon this the gold and copper will be thrown to the bottom of the vessel with the silver. Let stand till fully settled, then pour off the liquid carefully, and refill with boiling rain water as before. Continue to repeat this operation until the precipitate is thoroughly washed; or, in other words, fill up, let settle, and pour off so long as the accumulation at the bottom of the vessel is acid to the taste.

You now have about an eighteen carat chloride of gold. Add to it an ounce and an eighth cyanuret potassa, and one quart rain water—the latter heated to the boiling point. Shake up well, then let stand about twenty-four hours and it will be ready for use.

Some use platina as an alloy instead of silver, under the impression that plating done with it is harder. I have used both, but never could see much difference.

Solution for a darker colored plate to imitate Guinea gold may be made by adding to the above one ounce of dragon's blood and five grains iodide of iron.

If you desire an alloyed plate, proceed as first directed, without the silver or copper, and with an ounce and a half of sulphuret potassa in place of the iron, borax and salt.

TO MAKE SILVER SOLUTION FOR ELECTRO-PLATING.

Put together into a glass vessel, one ounce good silver, made thin and cut into strips; two ounces best nitric acid and half an ounce pure rain water. If solution does not begin at once, add a little more water—continue to add a very little at a time till it does. In the event it starts off well, but stops before the silver is fully dissolved, you may generally start it up again all right by adding a little more water.

When solution is entirely effected, add one quart of warm rain water and a large tablespoonful of table salt. Shake well and let settle, then proceed to pour off and wash through other waters as in the case of the gold preparation. When no longer acid to the taste, put in an ounce and an eighth cyanuret potassa and a quart pure rain water; after standing about twenty-four hours it will be ready for use.

TO PLATE WITH A BATTERY.

If the plate is to be gold use the gold solution for electro-plating; if silver, use the silver solution. Prepare the article to be plated by immersing it for several minutes in a strong ley made of potash and rain water, polishing off thoroughly at the end of the time with a soft brush and prepared chalk. Care should be taken not to let the fingers come in contact with the article while polishing, as that has a tendency to prevent the plate from adhering—it should be held in two or three thicknesses of tissue paper.

Attach the article, when thoroughly cleansed, to the positive pole of your battery, then affix a piece of gold or silver, as the case may be, to the negative pole, and immerse both into the solution in such a way as not to hang in contact with each other.

After the article has been exposed to the action of the battery about ten minutes, take it out and wash or polish over with a thick mixture of water and prepared chalk or jeweler's rouge. If, in the operation, you find places where the plating seems inclined to peel of, or when it has not taken well, mix a little of the plating solution with prepared chalk or rouge, and rub the defective part thoroughly with it. This will be likely to set all right.

Govern your time of exposing the article to the battery by the desired thickness of the plate. During the time it should be taken out and polished up as just directed about every ten minutes, or as often at least as there is an indication of a growing darkness on any part of its surface. When done, finish with the burnisher on prepared chalk and chamois skin, as best suits your taste and convenience.

In case the article to be plated is iron, steel, lead, pewter, or block tin, you must, after first cleansing with the ley and chalk, prepare it by applying with a soft brush—a camel's hair pencil is best suited—a solution made of the following articles in the proportion named :—Nitric acid, half an ounce; muriatic acid, one third of an ounce; sulphuric acid, one ninth of an ounce; muriatic of potash, one seventh of an ounce; sulphate of iron, one fourth of an ounce; sulphuric ether, one fifth of an ounce, and as much sheet zinc as it will dissolve. This prepares a foundation, without which the plate would fail to take well, if at all.

TO PLATE WITHOUT A BATTERY.

Prepare the article same as to plate with a battery, then attach to a strip of sheet zinc and suspend in the gold or silver solution for electro-plating as the case may be. The zinc is usually passed around the object to be plated, though this is of no particular importance, all that is necessary is to have the metals in actual contact. Observe the same rules as laid down in the directions for plating with a battery. If the article being plated has the strip of zinc touching much of its surface, it may be well to change the place of contact at every polishing.

You will find this mode of plating but little inferior to that of plating with a battery. It is more employed now, perhaps, than any other.

TO MAKE GOLD AMALGAM.

Eight parts of gold and one of mercury are formed into an amalgam for plating by rendering the gold into thin plates, making it red hot and then putting it into the mercury while the latter is also heated to ebullition. The gold immediately disappears in combination with the mercury, after which the mixture may be turned into water to cool. It is then ready for use.

TO PLATE WITH GOLD AMALGAM.

Gold amalgam is chiefly used as a plating for silver, copper or brass. The article to be plated is washed over with diluted nitric acid or potash ley and prepared chalk, to remove any tarnish or rust that might prevent the amalgam from adhering. After having been polished perfectly bright the amalgam is applied as evenly as possible, usually with a fine scratch brush. It is then set upon a grate over a charcoal fire, or placed into an oven and heated to that degree at which mercury exhales. The gold, when the mercury has evaporated, presents a dull yellow color. Cover it with a coating of pulverized nitre and alum in equal parts, mixed to a paste with water, and heat again till *it* is thoroughly melted, then plunge into water. Burnish up with a steel or bloodstone burnisher.

TO MAKE AND APPLY GOLD PLATING SOLUTION.

Dissolve half an ounce of gold amalgam in one ounce of nitro-muriatic acid. Add two ounces of alcohol, and then, having brightened the article in the usual way, apply the solution with a soft brush. Rinse and dry in saw-dust, or with tissue paper, and polish up with chamois skin.

TO MAKE AND APPLY GOLD PLATING POWDERS.

Prepare a chloride of gold the same as for plating with a

battery. Add to it when thoroughly washed out, cyanuret potassa in the proportion of two ounces to five pennyweights of gold. Pour in a pint of clean rain water, shake up well and then let stand till the chloride is dissolved. Add then one pound of prepared Spanish whiting and let evaporate in the open air till dry, after which put away in a tight vessel for use. To apply it you prepare the article in the usual way, and having made the powder into a paste with water, rub it upon the surface with a piece of chamois skin or cotton flannel.

An old mode of making a gold plating powder was to dip clean linen rags into solution prepared as in the second article preceding this, and having dried, to fire and burn them into ashes. The ashes formed the powder, and were to be applied as above.

TO MAKE AND APPLY SILVER PLATING SOLUTION.

Put together in a glass vessel one ounce nitrate of silver, two ounces cyanuret potassa, four ounces prepared Spanish whiting and ten ounces pure rain water. Cleanse the article to be plated as per preceding directions, and apply with a soft brush. Finish with the chamois skin or burnisher.

TO MAKE AND APPLY SILVER PLATING POWDER.

Dissolve silver in nitric acid by the aid of heat; put some pieces of copper into the solution to precipitate the silver; wash the acid out in the usual way; then with fifteen grains of it mix two drachms of tartar, two drachms of table salt and half a drachm of pulverized alum. Brighten the article to be plated with ley and prepared chalk, and rub on the mixture. When it has assumed a white appearance expose to heat as in the case of plating with gold amalgam, then polish up with the burnisher or soft leather.

TO SILVER IVORY.

Immerse the ivory in a weak solution of nitrate of silver till it takes upon itself a bright yellow color; take it then from the solution and expose, under water, to the rays of the sun. In two or three hours it will become black; but on taking it out of the water and rubbing it, the blackness will change to a beautiful silvering.

TO SILVER GLASS GLOBES.

Take equal parts of tin and lead, and melt them together; add while they are in fusion two parts of bismuth and two parts of mercury. Take from the fire, and so soon as cool enough for the glass to bear it, pour into the globe and move slowly so that the amalgam will pass over every part of its interior. A thin film will be left at every point of contact.

CHAPTER XI.

MISCELLANEOUS RECIPES.

TO FROST WATCH MOVEMENTS.

Sink that part of the article to be frosted for a short time in a compound of nitric acid, muriatic acid and table salt—one ounce of each. On removing from the acid, place it in a shallow vessel containing enough sour beer to merely cover it, then with a fine scratch brush scour thoroughly, letting it remain under the beer during the operation. Next wash off, first in pure water and then in alcohol. Gild or silver in accordance with any recipe in the chapter on plating.

TO MAKE CLEANSING SOLUTION FOR BRASS.

Put together two ounces sulphuric acid, an ounce and a half nitric acid, one dram saltpetre and two ounces rain water. Let stand for a few hours, and apply by passing the article in and out quickly, and then washing off thoroughly with clean rain water. Old discolored brass chains treated in this way will look equally as well as when new. The usual method of drying is in sawdust.

TO MAKE AND APPLY SOLUTION FOR FROSTING SILVER ARTICLES.

Put one dram of sulphuric acid into four ounces of rain water. Heat the solution and sink the silver in it till frosted as desired then wash clean and dry in sawdust. Half a dram of acid to four ounces water makes a good solution for whitening silver articles.

POLISHING POWDER FOR GOLD ARTICLES.

Dr. W. Hofman has analyzed a polishing powder sold by gold workers in Germany, which always commands a very high price, and hence, it may be inferred, is well adapted for the purpose. He found it to be a very simple composition, being a mixture of about 70 per cent. of sesquioxide of iron and 30 per cent. of sal-ammoniac. To prepare it, protochloride of iron, prepared by dissolving iron in hydrochloric acid, is treated with liquid ammonia until a precipitate is no longer formed. The precipitate is collected on a filter, and without washing, is dried at such a temperature that the adhering sal-ammoniac shall not be volatilized. The peroxide of iron precipitate at first becomes charged with sesquioxide.

TO REMOVE TARNISH FROM ELECTRO-PLATED GOODS.

Make a solution of half a pound cyanuret potassa in two gallons rain water. Immerse the article till the tarnish has disappeared, then rinse off carefully in three or four waters, and dry in sawdust.

TO MAKE RED WATCH HANDS.

Mix together and hold over a lamp, until formed into a paste, one ounce carmine, one ounce muriate of silver and half an ounce tinner's japan. Apply to the watch hands, lay them on a copper plate, face up, and then hold the plate over your spirit lamp till they assume the color you desire.

TO GIVE PLASTER FIGURES THE APPEARANCE OF BRONZE.

Make a preparation of palm soap, five ounces; sulphate of copper, one and a half ounces, and sulphate of iron, one and a half ounces. Dissolve the soap in rain water in one vessel and the sulphates in another. Put together and let settle, then pour off the water. Dry the precipitate, and apply to the figure by mixing as a paint with linseed oil and turpentine.

TO ETCH ON IVORY,

Cover the ivory to be etched with a thin coating of bees-wax, then trace the figure you desire to present through the wax. Pour over it a strong solution of nitrate of silver. Let remain a sufficient length of time, then remove it, with the wax, by washing in warm water. The design will be left in dark lines on the ivory.

TO ENAMEL GOLD OR SILVER.

Take half a pennyweight of silver, two pennyweights and a half of copper, three pennyweights and a half of lead and two pennyweights and a half of muriate of ammonia. Melt together and pour into a crucible with twice as much pulverized sulphur; the crucible is then to be immediately covered that the sulphur may not take fire, and the mixture is to be calcined over a smelting fire until the superfluous sulphur is burned away. The compound is then to be coarsely pounded, and with a solution of muriate of ammonia to be formed into a paste, which is to be placed upon the article it is designed to enamel. The article must then be held over a spirit lamp till the compound upon it melts and flows. After this it may be smoothed and polished up in safety. This makes the black enamel now so much used on jewelry.

TO DESTROY THE EFFECT OF ACIDS ON CLOTHES.

Dampen as soon as possible after exposure to the acid with spirits ammonia. It will destroy the effect immediately.

TO WASH SILVER WARE.

Never use a particle of soap on your silver ware, as it dulls the lustre, giving the article more the appearance of pewter than silver. When it wants cleaning rub it with a piece of soft leather and prepared chalk, the latter made into a kind of paste with pure water. I say pure water, for the reason that water not pure might contain gritty particles.

TO CLEANSE BRUSHES.

The best method of cleansing watchmakers' and jewelers' brushes is to wash them out in strong soda water. When the packs are wood you must favor that part as much as possible, for being glued the water may injure them.

TO CUT GLASS ROUND OR OVAL WITHOUT A DIAMOND.

Scratch the glass around the shape you desire with the corner of a file or graver; then, having bent a piece of wire to the same shape, heat it red hot and lay it upon the, scratch, sink the glass into cold water just deep enough for the water to come almost on a level with its upper surface. It will rarely ever fail to break perfectly true.

TO RE-BLACK CLOCK HANDS.

Use asphaltum varnish. One coat will make old rusty hands look as good as new, and it dries in a very few minutes.

GLOSSARY.

ARBOR.—An axle which turns upon itself by means of its pivots. Some watchmakers apply the term only to the post on which the key is placed to wind the watch, and to the rod passing through the cannon.

ANCHOR.—A piece of the escapement used in clocks and lever watches.

ANCHOR ESCAPEMENT WATCH.—A detached lever is often called an anchor escapement.

BARREL.—That piece of the watch which contains the main spring.

BRIDGE.—A piece secured to the plate, in which a pivot works, as in the case of skeleton levers.

BALANCE.—A wheel which moves back and forth in obedience to the adverse action of the lever and hair spring.

BEAT.—Each "tick" of the watch is called a beat.

CLICK.—A small lever which works into a ratchet and prevents the sudden recoil of the mainspring when the watch is wound up.

CENTRE WHEEL.—The large wheel immediately in the centre of the watch.

CHICK.—A small pin; usually those which hold the bridges in position.

CYLINDER.—The hollow piece which checks the onward motion of the scape wheel in cylinder escapement watches.

CANNON.—The steel piece which comes up through the dial, and around which the hour wheel revolves. In English and American levers the minute hand is fastened upon it.

COMMON PINION.—The pinion at the lower end of the cannon, which moves the minute wheel.

COCK.—Bridge over the balance.

COLLET.—A small ring fitting on the balance staff or arbor, and holding the inside end of the hair spring. The rings into which jewels are sometimes set are also called collets.

DIAL.—The face of the watch or clock.

DIAL WHEELS.—Those working between the dial and pillar plate.

DEPTHING TOOL.—An instrument used for finding the proper location of pivot holes.

ESCAPEMENT.—Those pieces in the watch or clock which work together and regulate the velocity of the time train.

ELECTRO PLATING—Plating through the aid of electricity. Formerly it was only done with a battery, but recent discoveries enable us to make a very good electro-plate without a battery.

FUSEE.—The cone-shaped wheel upon which the chain works.

FOURTH WHEEL.—The wheel which, in ordinary watches, works into the scape wheel.

FOURTH WHEEL SECOND.—A watch carrying a second hand on the pivot of its fourth wheel.

FORK.—That part of the lever into which the ruby pin plays.

FLY.—An arbor carrying two wings for the purpose of meeting with atmospheric resistance, and thus regulating the motion of striking trains in clocks.

GUARD POINT.—The wedge-shaped elevation immediately back of the fork in detached levers.

GEARING.—The action of the teeth of one wheel upon those of another wheel or pinion.

HOROLOGY.—That branch of science which treats of the principles and construction of machines for measuring time.

HOROLOGIST.—One who interests himself in the science of horology. A constructor or repairer of machines for measuring time. Strictly applicable to the American watch maker, owing to the fact that he works on all manner of time machines.

HOUR WHEEL.—The wheel working around the cannon, and upon which the hour hand is fastened.

INDEX.—Hand.

JEWEL.—The stone or glass settings through which or against which the pivots work ; also the settings in the pallets and the roller.

LEAVES.—Teeth or cogs of a pinion.

LEVER.—A horizontal bar upon which the pallets are secured, and which conducts the effect of the motive power from the train to the balance.

MINUTE WHEEL.—A name generally given to that dial wheel which is driven by the cannon pinion.

MOVEMENT.—The interior works of the clock or watch, independent of case.

PALLETS.—The jeweled piece of the lever watch which works into the teeth of the scape wheel.

PIVOT.—The end of an arbor turned very small to avoid friction.

PINION.—A small leaved wheel.

PILLARS.—Posts which in plate watches hold the plates the proper distance apart for the working of the train or trains between them.

PILLAR PLATE.—Usually the bottom plate of a watch, but European watchmakers generally call both pillar plates, distinguishing them as the upper and lower.

PUTTING UP.—Setting the different parts of a clock or watch into their proper places.

PIVOT WOOD.—A tough wood employed by watchmakers in cleaning out pivot holes. It is sometimes called peg wood. A scape wheel is sometimes called a ratchet in Europe.

RATCHET.—A steel wheel into which the click works.

RUBY PIN.—A small glass or stone pin which works in connection with the lever.

ROLLER.—The circular plate into which the ruby pin is set. It is often called the ruby pin table.

ROLL PLATE.—The best grade of plated jewelry.

STOP WORKS.—A mechanism to prevent the watch from being wound up too far.

STAFF.—A name generally applied to the balance arbor of lever watches; also to the arbor passing through the pallets.

SCAPE WHEEL.—The last wheel of the train.

SECOND WATCHES.—Watch with a second hand.

SCRATCH BRUSH.—A brush made of fine brass wire.

TEETH —Cogs by which the motion of one wheel is communicated to another.

TRAIN.—A collection of wheels so arranged that the moving power applied to the first wheel is freely communicated to them all.

THIRD WHEEL.—The wheel into whose pinion the centre wheel works.

TAKING DOWN.—Taking apart the different pieces of a clock or watch.

WHEEL BED.—A bed turned out in the plate of a watch to receive a wheel.

www.ingramcontent.com/pod-product-compliance
Lightning Source LLC
Chambersburg PA
CBHW031748090426
42739CB00008B/924